YOUR KNOWLEDGE HAS VALUE

Md. Siddiqur Rahman

Introduction to Remote Sensing

The Theory

GRIN Verlag

Bibliografische Information der Deutschen Nationalbibliothek:

Die Deutsche Bibliothek verzeichnet diese Publikation in der Deutschen National-
bibliografie; detaillierte bibliografische Daten sind im Internet über http://dnb.d-
nb.de/ abrufbar.

Imprint:

Copyright © 2011 GRIN Verlag GmbH
Druck und Bindung: Books on Demand GmbH, Norderstedt Germany
ISBN: 978-3-656-20374-2

This book at GRIN:

http://www.grin.com/en/e-book/194960/introduction-to-remote-sensing

GRIN - Your knowledge has value

Der GRIN Verlag publiziert seit 1998 wissenschaftliche Arbeiten von Studenten, Hochschullehrern und anderen Akademikern als eBook und gedrucktes Buch. Die Verlagswebsite www.grin.com ist die ideale Plattform zur Veröffentlichung von Hausarbeiten, Abschlussarbeiten, wissenschaftlichen Aufsätzen, Dissertationen und Fachbüchern.

Visit us on the internet:

http://www.grin.com/

http://www.facebook.com/grincom

http://www.twitter.com/grin_com

Introduction to Remote Sensing

A report submitted to Mr. A. Z. M. Zahidul Islam

Principal Scientific Officer

Head of the Division

Division of Water Resources

Bangladesh Space Research and Remote Sensing Organization (SPARRSO)

Agargaon, Sher-E-Bangla Nagar, Dhaka 1207, Bangladesh

Submitted by

Md. Siddiqur Rahman

Scientific Officer

Climate Change Research and Impact Study (CRAIST) Project

Bangladesh Space Research and Remote Sensing Organization (SPARRSO)

September, 2011

1. Definition of Remote Sensing (RS)

Remote sensing is the acquisition of information about an object or phenomenon, without making physical contact with the object (Wikipedia). Remote sensing is a generic term which describes the action of obtaining information about an object with a sensor which is physically separated from the object. It primarily concerned with deriving information about the earth's surface using an elevated platform (Harrison and Jupp, 1989).

2. Brief history of RS

Brief history of RS is adapted from Reeves et al. (1975) and Verstappen (1977).

o In 1783, first balooning by Marquis d' Arlandes and Pilatre d' Rosier in Paris.

o First photographs taken in 1839, next year, in 1840, first topographic maps were prepared.

o In 1858, Parisian photographer and baloonist G.F. Tornachon (known as Nadar) made the first aerial photographs. Same year, Laussedat developed a mathematical analysis for converting overlapping perspective views into orthographic projections on any plane.

o In 1860 and 1886, first aerial photographs were taken in Boston (USA) and Russia respectively from a free baloon.

o First photo interpretation during 1967-68-69 in USA for geological survey.

o In 1882 and 1895, kites were used in England and Western hemisphere to take photos.

o In 1903, carrier Pigeons were used to take photos by J. Neubronner at 30 second intervals.

o At the end of nineteenth century, Rockets were first used to take photos by the Germans.

o At the beginning of twentieth century, improved aircraft, camera system and photographic illusions have gradually resulted in ever improved RS technology.

o In 1946, hyper altitude photos taken by earth observatory satellites.

o In 1960, the term remote sensing was coined in lieu of aerial photography.

3. RS system

A system denotes the total appearance of all components, their interactions in doing a particular job. Remote sensing system comprises all the available components (both hard and soft wares), the online server, their interactions and interrelations in doing a particular job like image processing from satellite data. RSS comprises following components:

Platforms: The Satellite carrying the remote sensing device.

Sensor: The remote sensing device recording wavelengths of energy.

Computer system: The total computer hardware and software system.

Server: The server of the software development company that links with server stations.

Institutional setups: Cartography, image processing, other infrastructures etc.

Energy source: There must be an energy source for viewing the surface feature may be the Sun.

Image interpretation: The system of image interpretation is done by analyzing the captured image.

4. Interactions of EME with earth surface

Electro-magnetic energy (EME) interactions with atmosphere and with the earth surface play a vital role in remote sensing. Energy interactions with the atmosphere dictate the spectral regions through which only we can do the remote sensing which are known as Spectral windows (the spectral regions where atmosphere is more or less transparent).

The total incident energy will interact with earth's surface materials in three ways. These are: Absorption, Transmission, and Reflection. Absorption (A) occurs when radiation (energy) is absorbed into the target while transmission (T) occurs when radiation passes through a target. Reflection (R) occurs when radiation "bounces" off the target and is redirected.

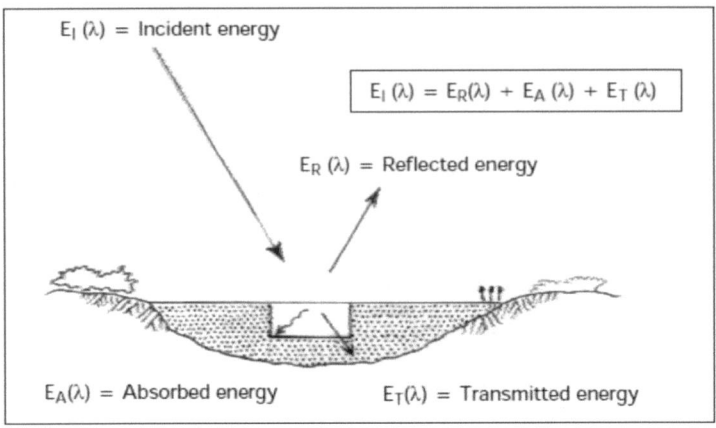

$E_I (\lambda)$ = Incident energy

$$E_I (\lambda) = E_R(\lambda) + E_A (\lambda) + E_T (\lambda)$$

$E_R (\lambda)$ = Reflected energy

$E_A(\lambda)$ = Absorbed energy $E_T(\lambda)$ = Transmitted energy

Figure 1 Interaction of Energy with the earth's surface (Liliesand & Kiefer, 1993).

5. Platforms and Sensors used in RS

Platforms

Platform means a device or base upon which remote sensing devices may be operated in variety of locations. RS technology used various types of platforms (adapted from Reeves et al., 1975).

Baloons – free baloons, powered baloons, tethered baloons.

Aircraft – Conventional aircraft: 1. Low altitude, 9 km, propeller driven. Eg. Cessna, Skymaster, Mohawk, Jet, Boeing. 2. High altitude, 15 km. Eg. Rockwell, Douglas, lockheed. Unconventional aircraft: Helicopters, Drone, Dirigibles, Sailplane.

Space craft – Unmanned: short lived, 150-500 km altitude, sun synchronous. Eg. Nimbus, TIROS, ERTS, ATS, SMS, Ranger, LOP, Viking, Pioneer. Manned: Mercury, GEmini, Apollo, Skylab, Space Shuttles etc.

Sensors

The device that is responsible for remote sensing (satellite) data acquisition (photo/image capture) is termed as Sensor. There are two types of remote sensing sensors, viz. photographic sensors and electromagnetic sensors.

Table 1 Sensors flown on satellites with their spatial resolution (Szekielda, 1986)

Satellites	Sensor	Spatial resolution
Nimbus I-VII	High Resolution Infrared Radiometer (HRIR)	8 km
NOAA 1-4	Scanning Radiometer (SR)	7 km
TIROS -N	Advance Very High Resolution Radiometer (AVHRR)	1 km
LANDSAT 1-2	Multispectral Scanner (MS)	70 m
Soyuz-22	Multispectral Camera (MKF-6)	10 m

6. Radiometric and Geometric characteristics of remotely sensed images

Radiometric characteristics Radiometric characteristics of remotely sensed images incorporated with different technical terms and physical units.

o Radiant energy: the energy carried by electro-magnetic radiation (J).

o Radiant flux: radiant energy transmitted as a radial direction per unit time (W).

o Radiant intensity: radiant flux radiated from a point source/ unit solid angle (Wsr^{-1}).

o Irradiance: radiant flux incident upon a surface per unit area (Wm^{-2}).

o Radiant emittance: radiant flux radiated from a surface per unit area (Wm^{-2}).

o Radiance: radiant intensity per unit projected area in a radial direction ($Wm^{-2} sg^{-1}$).

o Radiometric resolution: dynamic range or number of possible data files in each band.

o Film exposure of sensor: Film exposure at a point in an image directly related to the reflectance of the object imaged at that point.

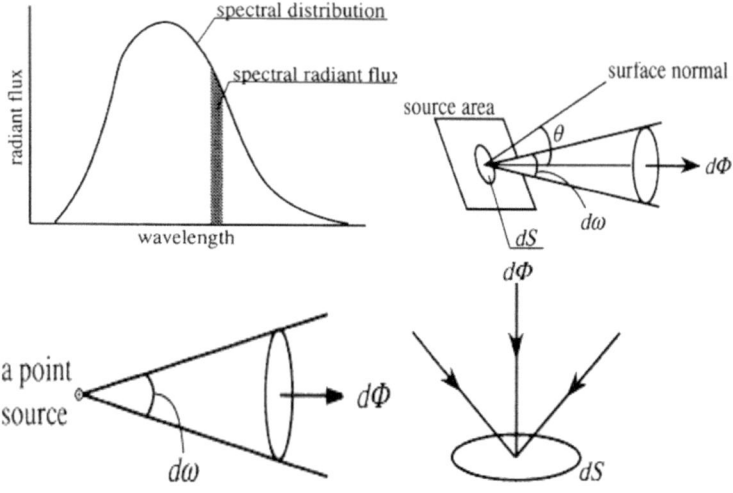

Figure 2 Radiometric factors those influence RS imagery data

Geometric Characteristics

Remote sensing data are data digitized by a process of sampling and quantization of the electro-magnetic energy, which detected by a sensor.

• IFOV (Instantaneous Field Of View) defined as the angle, which corresponds to the sampling unit. Information within an IFOV is represented by a pixel in the image plane.

• Extraneous Geometric effect (Adapted from Lillesand and Kiefer, 1979):

  o Exposure fall off: Because of fall off, a ground scene of spatially uniform reflectance does not produce spatially uniform exposure in the focal plane.

o   Oblique photograph: Oblique photographs of same area taken from different vantage points.

o   The 'sun-object-image' relationship of a scene is important. There are three angles- solar elevation, azimuth angle and viewing angle. A change in any one may change the apparent reflectance of objects.

o   Differential shading: Relief displacement of an RS data/image cause vertical features to be imaged slightly in side view as well as in top view.

7.   Information available on remotely sensed images

Locations and other characteristics of natural features and human activities on, above and beneath the earth's surface are recorded as information for RS. The raster (or grid-cell) data model has developed from aerial and satellite-imaging technology, which represents geographical objects as grid-cell structures known as pixels. The location of geographic objects or conditions is defined by the row and column position of the cells they occupy. The area of each cell defines a spatial resolution available (Aronoff, 1989). Information available on remotely sensed images may have three modes or dimensions, i.e. spatial, temporal, or thematic (Heywood et al., 1988):

• Spatial: The spatial dimension convey to the user information about the location of the feature observed. It refers size, shape, colour, tonal variance, shadow, texture, and pattern.

• Temporal: The temporal dimension provides a record of when the data were collected.

• Thematic/attribute: The thematic dimension shows the characteristic of a real world feature to which the data refer. In RS, thematic data are often referred as non-spatial, or attribute, data.

8. Processing of remotely sensed images

Humans are adept at visually interpreted data. Processing of remotely sensed data is important for the following causes:

➢ Contrasting colour of images makes them easily understood by end users.

➢ Human interpretations are highly subjective, hence, not perfectly repeatable. Conversely, results generated by computer--even when erroneous--are usually repeatable.

➢ When very large amounts of data are, the computer may be better suited to managing the large body of detailed (and tedious) data.

Georeferencing

It refers to the process of assigning map coordinates to image data. The remote sensing data may already be projected onto the desired plane (resample), but not yet referenced to the proper coordinate system. It provides a coordinate system to an (ortho-rectified) image or map. More specifically, Geo-referencing refers to the location of a layer or coverage in space defined by the coordinate referencing system. It may involve shifting, rotating, scaling, skewing, and in some cases warping, rubber sheeting, or orthorectifying the data.

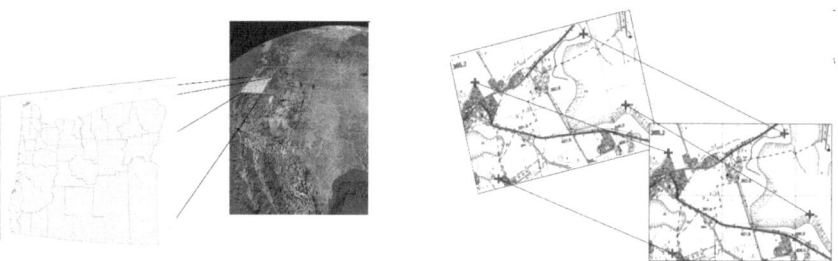

Figure 3 Georeferncing of real world features to map surfaces.

Classification

The process of sorting or arranging entities into groups or categories; on a map, the process of representing members of a group by the same symbol, usually defined in a legend is termed as image classification of RS data. It is the process of assigning the pixels of a continuous raster image to discrete categories. Classification system is a set of target classes. The purpose of such a scheme is to provide a framework for organizing and categorizing the information that may be extracted from the data. There are two types of classification: Supervised and Unsupervised classification.

Analysis

A systematic examination of a problem or complex entity in order to provide new information from what is already known. Geographic bounding area within which spatial analysis will occur. The bounding area is set by defining the x,y coordinates of opposite corners.

9. Spectral characteristics of remotely sensed data (Landsat TM)

- Bandwidth
  - o Range of wavelengths (colors) detected by a particular band
- Band placement
  - o The portion of the electromagnetic spectrum detected by a particular band
  - o Defined by the low and high wavelengths of the range or the by center of the range
- Number of bands
  - o The number of bands imaged by the sensor

- o Often grouped as panchromatic (single band), multispectral (more than one band), or hyperspectral (usually over 100 bands)

Landsat TM Band spectral characteristics

The TM bands were chosen after years of analysis of MSS and other remote sensing data for their value in water penetration, discriminating vegetation types and vigour, plant and soil moisture measurements, differentiation of clouds, snow, and ice identification hydrothermal variation in certain rock types and, atmospheric effects determine the placement of bands in the spectrum.

Table 2 Spectral range of bands and spatial resolution for the TM sensor

Bands name	Spectral range (µm)	Bands	Resolution
Band 1	0.45-0.52	blue	30
Band 2	0.52-0.60	green	30
Band 3	0.63-0.69	red	30
Band 4	0.76-0.90	near infrared	30
Band 5	1.55-1.75	mid infrared	30
Band 6	10.40-12.50	thermal infrared	120
Band 7	2.08-2.35	mid infrared	30

All TM bands are quantized as 8 bit data.

Band 1: Provides increased penetration of water bodies as well as supporting analyses of land use, soil, and vegetation characteristics. Wavelengths below 0.45 m are substantially influenced by atmospheric scattering and absorption.

Band 2: This band corresponds to the green reflectance of healthy vegetation and is spanning the region between the blue and red chlorophyll absorption bands.

Band 3: This red chlorophyll absorption band of healthy green vegetation is one of the most important bands for vegetation discrimination. In addition, it is useful for soil-boundary and geological boundary mapping.

Band 4: It is useful for identification of vegetation types, and emphasizes soil-crop and land-water contrasts.

Band 5: This reflective-IR band is sensitive to turgidity - the amount of water in plants. It is also used to discriminate between clouds, snow, and ice which make it important in hydrologic research.

Band 6: This band measures the amount of infrared radiant flux (heat) emitted from surfaces in locating geothermal activity, thermal inertia mapping, vegetation classification, vegetation stress analysis, and in measuring soil moisture.

Band 7: This band is used to discriminate between geological rock formations. It is particularly effective in identifying zones of hydrothermal alteration in rocks.

References

Aronoff, S. 1989. Geographic Information Systems: A Management Perspective, WDL Publications, Ottawa, Canada.

Harrison, B.A. and Jupp, D.L.B., 1989. Introduction to Remotely Sensed Data. CSIrRo Publications. 314 Albert Street, East Melbourne, Australia. 141 pp.

Heywood I., Cornelius S. and Carver, S. 1988. An Introduction of Geographical Information Systems, Addison Wesley Longman, New York.

Lillesand T.M. and Kiefer R.W. 1979. Remote Senmsing and Image Interpretation. Ist Edition. John Wiley & Sons. Newyork. 612 pp.

Lillesand, T.M. and Kiefer, R.1993. Remote Sensing and Image Interpretation. Third Edition. John Villey, New York.

Reeves, R.G., Anson, A. and Landen, D., 1975. Manual of Remote Sensing. Volume 1. Theory, Instruments and Techniques. American Society of Photogrammetery, Falls Church, VErginia. 867 pp.

Szekielda, K.H., 1986. Satellite Remote Sensing for Resources Development. First Edition. Graham & Trotman Ltd., UK. 221 pp.

Verstappen, H.Th., 1977. Remote Sensing in GEomophology. Elsevier Scientific Publishing Co., Amsterdam, The Netherlands. 214 pp.